Spider Biters

Therese Shea

PowerKiDS press.
New York

Published in 2007 by The Rosen Publishing Group, Inc.
29 East 21st Street, New York, NY 10010

Book Design: Daniel Hosek

Photo Credits: Cover © pixelman/Shutterstock; p. 5 © Liv Falvey/Shutterstock; p. 7 © Robert Noonan; p. 9 © James P. Rowan; p. 11 © Ra'id Khalil/Shutterstock; p. 13 © Animals Animals; p. 13 (inset) © Dirk Herzog/Shutterstock; p. 15 © Robert and Linda Mitchell; p. 17 © Cindy Jenkins/Shutterstock; pp. 19, 21 © David Liebman; p. 22 © photobar/Shutterstock.

Library of Congress Cataloging-in-Publication Data

Shea, Therese.
Spider biters / Therese Shea.
p. cm. -- (Big bad biters)
Includes bibliographical references (p.).
ISBN-13: 978-1-4042-3521-3
ISBN-10: 1-4042-3521-3
1. Spiders--Juvenile literature. I. Title. II. Series: Shea, Therese. Big bad biters.
QL458.4.S534 2006
595.4'4--dc22
2006014647

Manufactured in the United States of America

Contents

A Riddle

What is small, has eight legs, and spins silk? A spider! Most people think spiders are just gross. Actually, they are amazing animals. Did you know that a spider eats only liquids? It sucks the liquid out of the body of its **prey**. Did you know that the blood in a spider's body helps move its legs? In this book, we'll find out much more about some of the coolest spiders in nature!

Spiders, like these orb weavers, cannot move without enough liquid in their bodies.

Spider Bodies

All spiders have two main body parts. The front part is made up of the head and chest. The spider's mouth, **fangs**, eyes, and legs are found on the front part. The rear part of the body contains silk **glands**. These glands make the silk that spiders use to build webs and catch prey.

Spiders have no bones. Their bodies are covered by tough skin and hairs that help them sense their surroundings.

Spiders outgrow their skin and shed it several times until they finish growing. This spider's old skin is above it.

So Many Eyes!

Most spiders have eight eyes. Others have six, four, or two eyes. Some spiders that live in dark places have no eyes at all! Hunting spiders can see well. This helps them run after their prey. Web-building spiders cannot see well. Their eyes can only tell dark from light. They use their silk webs to trap their prey.

This wolf spider has eight eyes. Wolf spiders are good hunters.

Scary Fangs

Spiders don't have teeth, and they don't chew. Instead, spiders have one small leglike body part on each side of their mouth. These are used to grab and crush their prey.

Spiders also have two body parts that act as claws and fangs. The spider stabs its prey with these fangs. **Venom** in the fangs stops the prey from moving or kills it. Then the spider sprays other juices that turn the inside of the prey's body into liquid. The spider drinks this liquid.

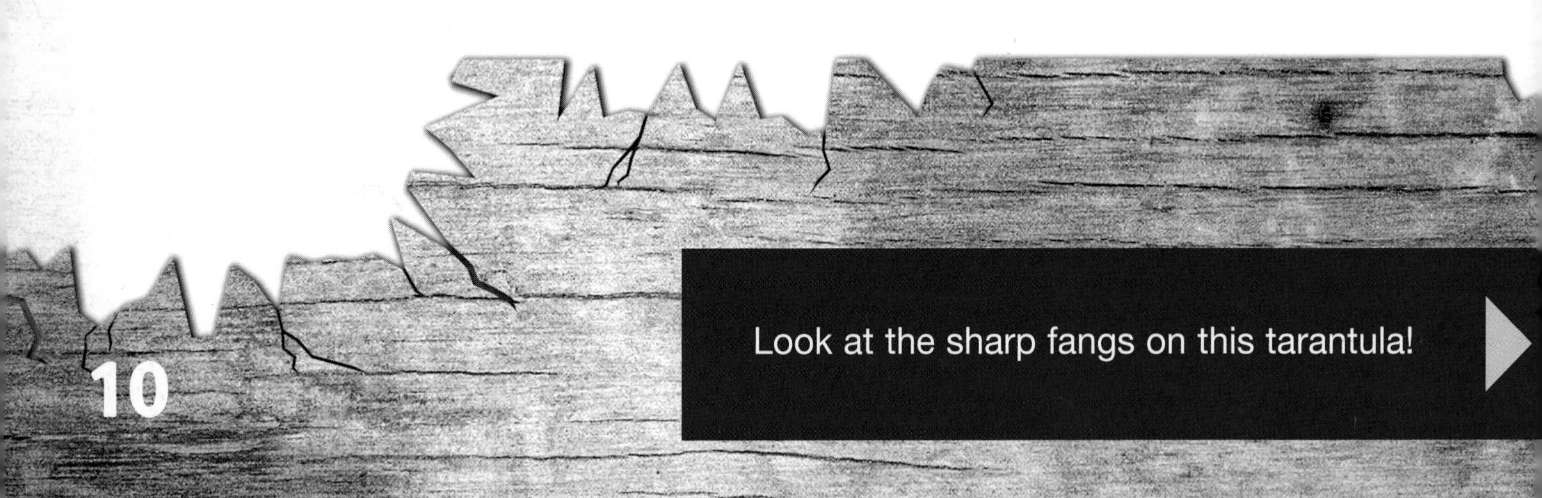

Look at the sharp fangs on this tarantula!

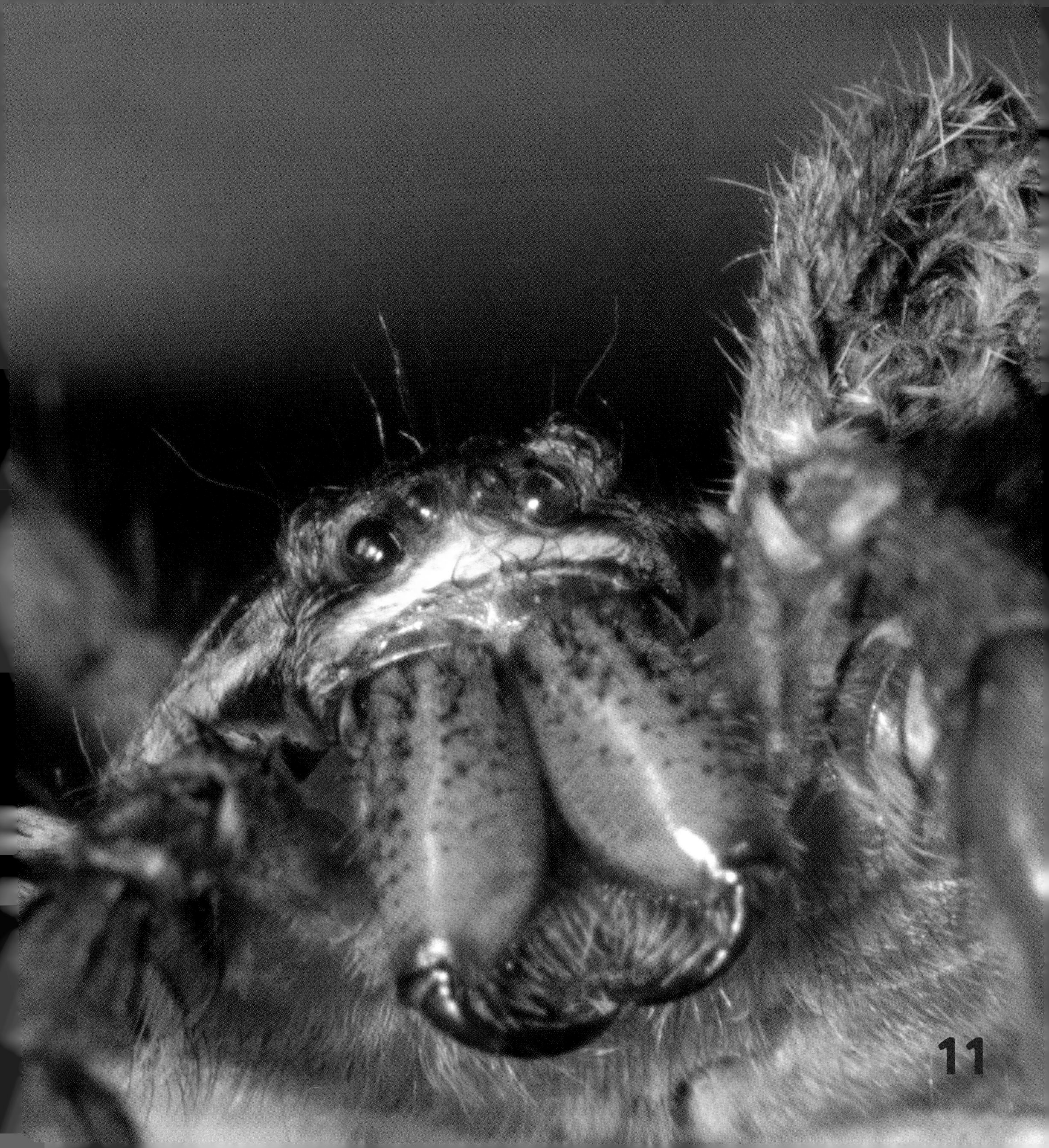

A Trail of Silk

Everywhere a spider travels, it trails a line of silk behind it. Spiders have silk glands in the rear part of their bodies. The glands make a liquid silk. The silk passes into body parts called **spinnerets**. Spinnerets spin the silk and send it through tubes to the outside of the body. When the silk reaches the air, it hardens into a thread. Some glands make silk that is dry when it hardens. Some make silk that is sticky.

Can you see the silk thread behind the jumping spider in the top picture?

spinnerets

So Much Silk!

Some spiders use silk to make sticky webs that catch prey. A spider may wrap its prey in silk so it can't escape. Spiders also make silk nests on leaves, underground, or in the middle of their webs.

Spiders use a **dragline** to escape from enemies. A spider can drop to the ground from its web using a dragline. It can also swing in the air from its dragline. When the enemy goes away, the spider climbs back up the dragline.

Female spiders, like this cobweb weaver, use a special silk to wrap their eggs.

Tarantulas

Tarantulas live in warm parts of the world. They hunt at night. When a tarantula feels the ground **vibrate**, it runs to grab its prey. Then it stands on its back legs and pushes its fangs into the prey's body. The venom stops the animal from moving, so the spider can eat it.

Tarantulas also use their fangs to scare their enemies. Their enemies include weasels, skunks, owls, snakes, and large wasps.

Small tarantulas eat mostly bugs. Giant jungle tarantulas eat frogs, lizards, snakes, and baby birds.

Orb Weavers

Orb weavers spin webs with amazing patterns to catch **insects**. Some orb weavers wait in the middle of their web for an insect. Others attach a special silk thread to the center of the web. They wait nearby for the thread to vibrate. The orb weaver then knows its prey is stuck to the web.

An orb weaver may wrap its prey in silk before biting it with its **venomous** fangs. The silk wrapping keeps prey, like wasps, from hurting the orb weaver.

There are about 3,000 kinds of orb weavers. They may look very different from each other.

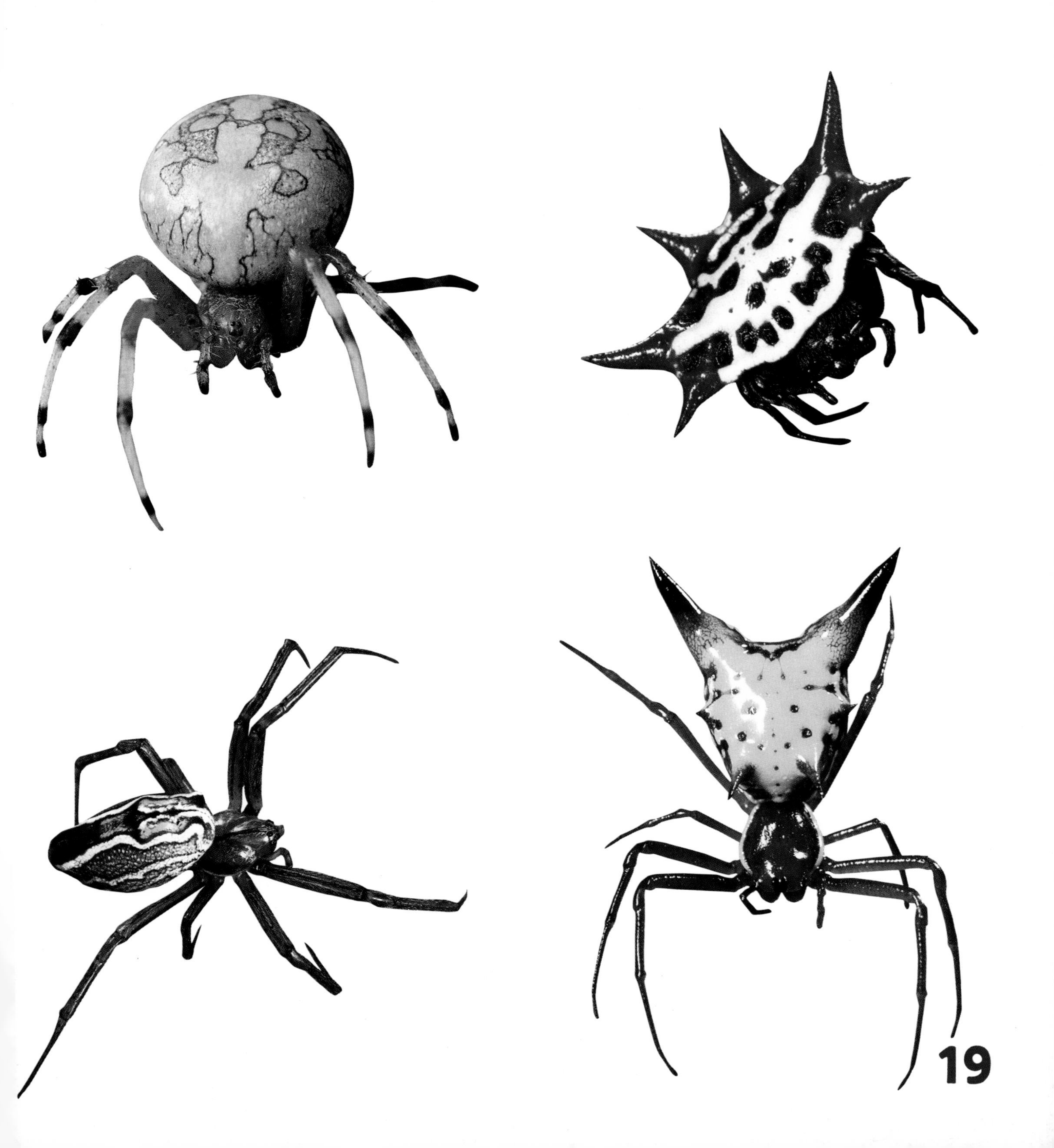

Fisher Spiders

Large and hairy fisher spiders live near water. They hunt small fish, insects, and tadpoles. Fisher spiders don't use webs to trap their prey. These spiders have large bodies and long, thin legs that allow them to walk on water. They can dive underwater to grab their prey with their legs. Like the venom of other spiders, their venom stops the animal from moving. Then the fisher spider drags it to land and eats it.

Fisher spiders can be as big as a child's hand!

How Spiders Help

We need spiders in our world. Spiders eat bugs that can bother people. People use spider silk in different ways, such as for clothes and fishing nets. Silk is the strongest fiber in nature!

Most spiders are only harmful to the small animals they eat. Some spiders can hurt humans, too. Black widow spiders and brown recluse spiders have bites than can harm a person. However, spiders only attack if they think they are in danger.

black widow

Glossary

dragline (DRAG-lyn) A line of silk that spiders use to attach themselves to something.

fang (FANG) A sharp, hollow, or grooved tooth that contains venom.

gland (GLAND) A body part that makes something to help the body function.

insect (IHN-sekt) A small animal without bones that has three main body parts, six legs, and usually has wings.

prey (PRAY) An animal that is hunted by another animal as food.

spinneret (spih-nuh-REHT) A spider's body part that releases silk.

venom (VEH-num) A poison passed by one animal into another by a bite or sting.

venomous (VEH-nuh-muhs) Having or making venom.

vibrate (VY-brayt) To move back and forth quickly.

Index

Web Sites

Due to the changing nature of Internet links, PowerKids Press has developed an online list of Web sites related to the subject of this book. This site is updated regularly. Please use this link to access the list:
http://www.powerkidslinks.com/biters/spiders/